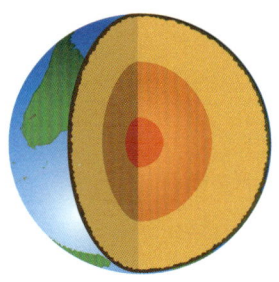

DeltaScience ContentReaders

Inside Earth

Contents

Preview the Book 2
What Is Earth Made Of? 3
    Earth's Layers 4
    Earth's Surface 5

How to Read Diagrams 8
How Does Earth Change? 9
    Earth's Plates 10
    How Mountains Form 12
    Studying Earth 14

Cause and Effect 16
What Are Volcanoes and Earthquakes? 17
    Volcanoes 18
    Earthquakes 21

Glossary 24

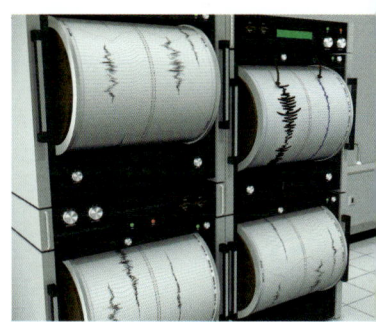

Build Reading Skills
Preview the Book

You read nonfiction books like this one to learn about new ideas. Be sure to look through, or *preview*, the book before you start to read.

First, look at the title, front cover, and table of contents. What do you guess you will read about? Think about what you already know about Earth.

Next, look through the book page by page. Read the headings and the words in bold type. Look at the pictures and captions. Notice that each new part of the book starts with a big photograph. What other special features do you find in the book?

Headings, captions, and other features of nonfiction books are like road signs. They can help you find your way through new information. Now you are ready to read!

What Is Earth Made Of?

MAKE A CONNECTION
Earth's land has steep cliffs, rolling hills, and many other features. Do you think the bottom of the ocean also has features like these? Why or why not?

FIND OUT ABOUT
- Earth's layers
- Earth's surface

VOCABULARY
crust, p. 4
mantle, p. 4
core, p. 5
landform, p. 5
continental shelf, p. 6
continental slope, p. 6
continental rise, p. 6
abyssal plain, p. 6

Earth's Layers

Earth is made of rock and metals. It has three main layers. They are the crust, the mantle, and the core.

The **crust** is Earth's outer layer. The crust is made of solid rock. Almost three-fourths of the crust is covered by oceans. A continent is a large area of the crust that can be seen above the ocean. Continental crust is thicker and lighter than the crust of the deep-ocean floor.

The **mantle** is the layer of rock under the crust. It is very hot there. The rock is under great pressure. The heat and pressure make some of the rock in the mantle flow very slowly.

Earth's three main layers are the crust, the mantle, and the core. ▼

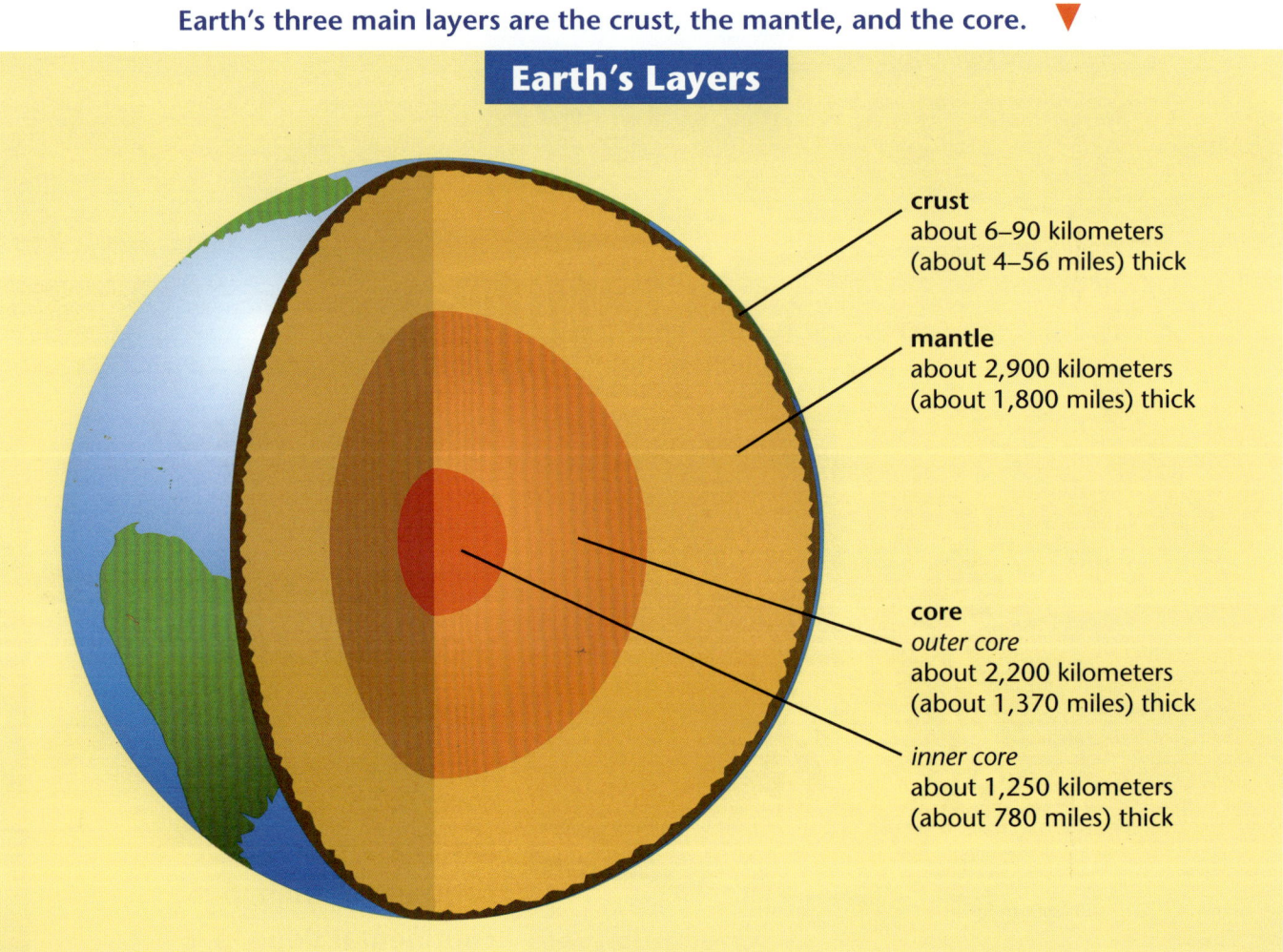

Earth's Layers

crust
about 6–90 kilometers
(about 4–56 miles) thick

mantle
about 2,900 kilometers
(about 1,800 miles) thick

core
outer core
about 2,200 kilometers
(about 1,370 miles) thick

inner core
about 1,250 kilometers
(about 780 miles) thick

The **core** is at the center of Earth. It is made mostly of metals. The *outer core* is liquid. The *inner core* is solid. The core is the hottest part of Earth.

 How are the layers of Earth different from one another?

Earth's Surface

Landforms are natural shapes, or features, on Earth's surface. Here are four kinds.
- *mountain* a landform that is much higher than the land around it
- *valley* a low area of land between mountains or hills
- *plain* a wide, flat area of land
- *plateau* a landform that is flat like a plain but is much higher than the land around it

A valley is one kind of landform. ▼

The ocean floor has many different landforms. ▶

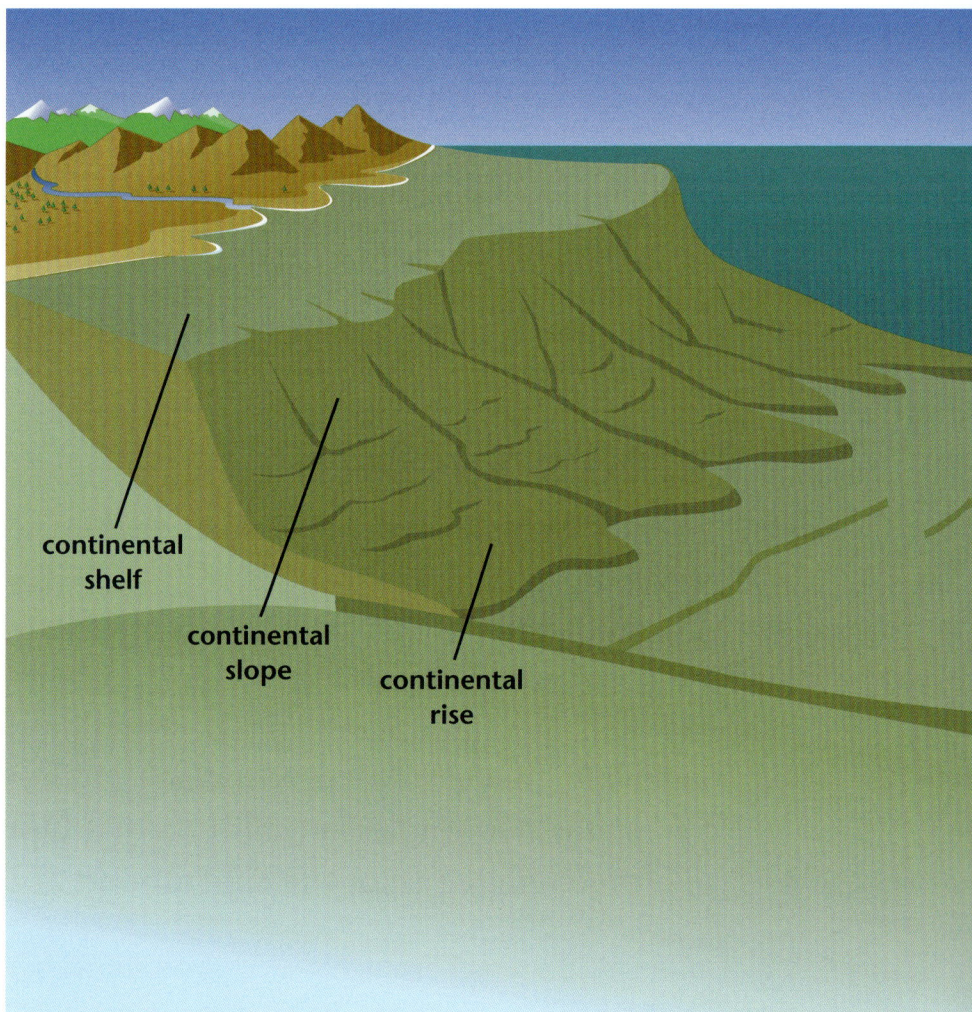

Like dry land, the ocean floor has landforms.

The **continental shelf** is the gently sloping underwater edge of a continent.

The **continental slope** is just beyond the continental shelf. There, the land drops off steeply.

The **continental rise** is at the bottom of the continental slope. There, the land starts to flatten out. The continental rise is made of small pieces of rock. The pieces fell down the continental slope.

An **abyssal plain** is a very wide, flat area of the deep-ocean floor.

The mid-ocean ridge is a chain of mountains that rises from the ocean floor. It is the longest chain of mountains on Earth. The mid-ocean ridge has a deep center valley.

Ocean Floor

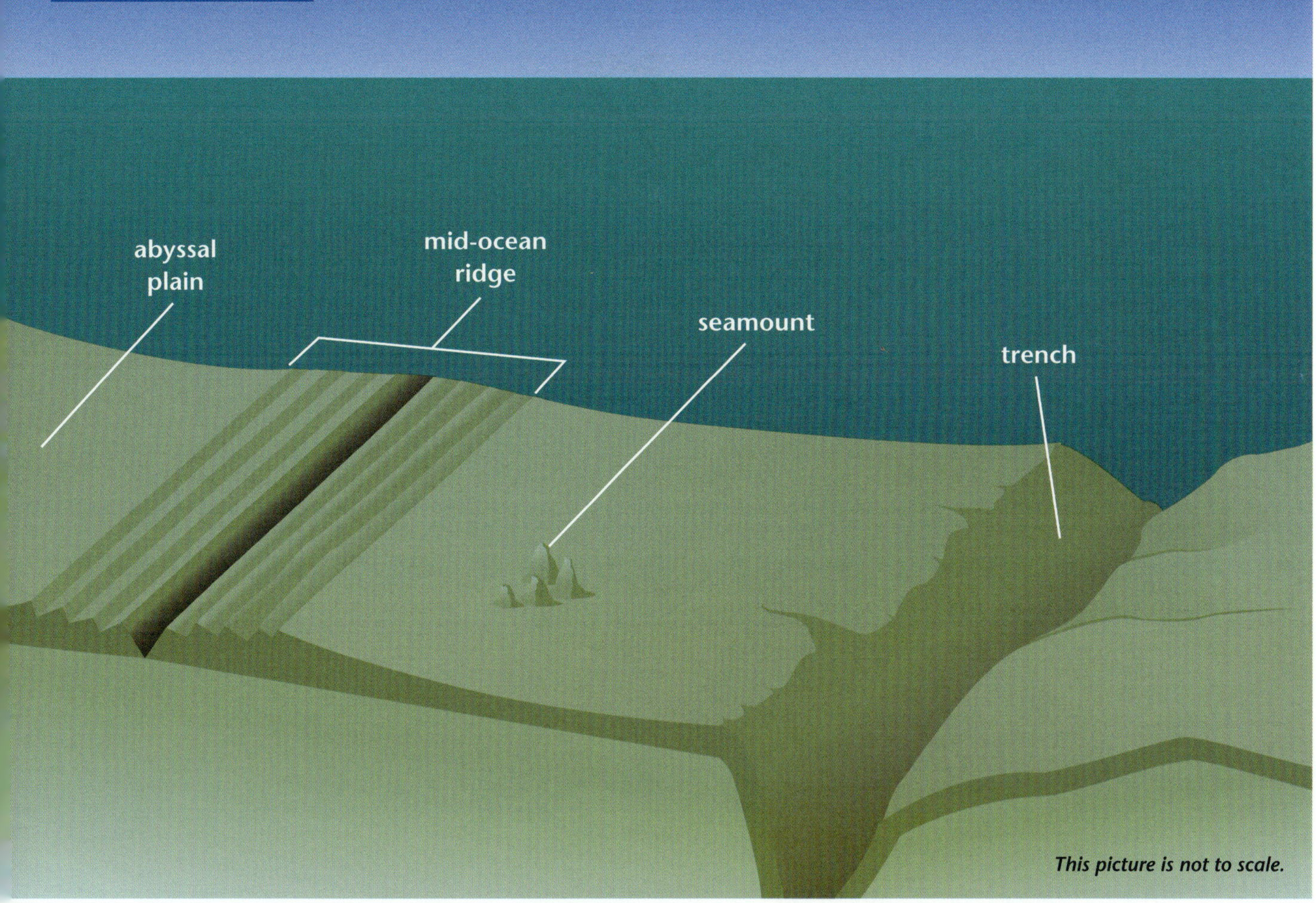

[Diagram labeled: abyssal plain, mid-ocean ridge, seamount, trench]

This picture is not to scale.

Hot, melted rock from Earth's mantle rises up into the valley. When the melted rock cools and hardens, it forms new crust.

A seamount is a volcano on the ocean floor. Some seamounts are tall enough to stick out of the water. They are called volcanic islands.

A trench is a very deep valley on the ocean floor.

 What is a landform? Name four kinds.

REFLECT ON READING
You previewed pictures, captions, and other book features before reading. Tell how one picture helped you better understand Earth's layers or Earth's surface.

APPLY SCIENCE CONCEPTS
With your class, make a poster showing pictures of some landforms on Earth. You should include mountains, valleys, plains, and plateaus. Label the landforms on the poster. Tell where on Earth they are found.

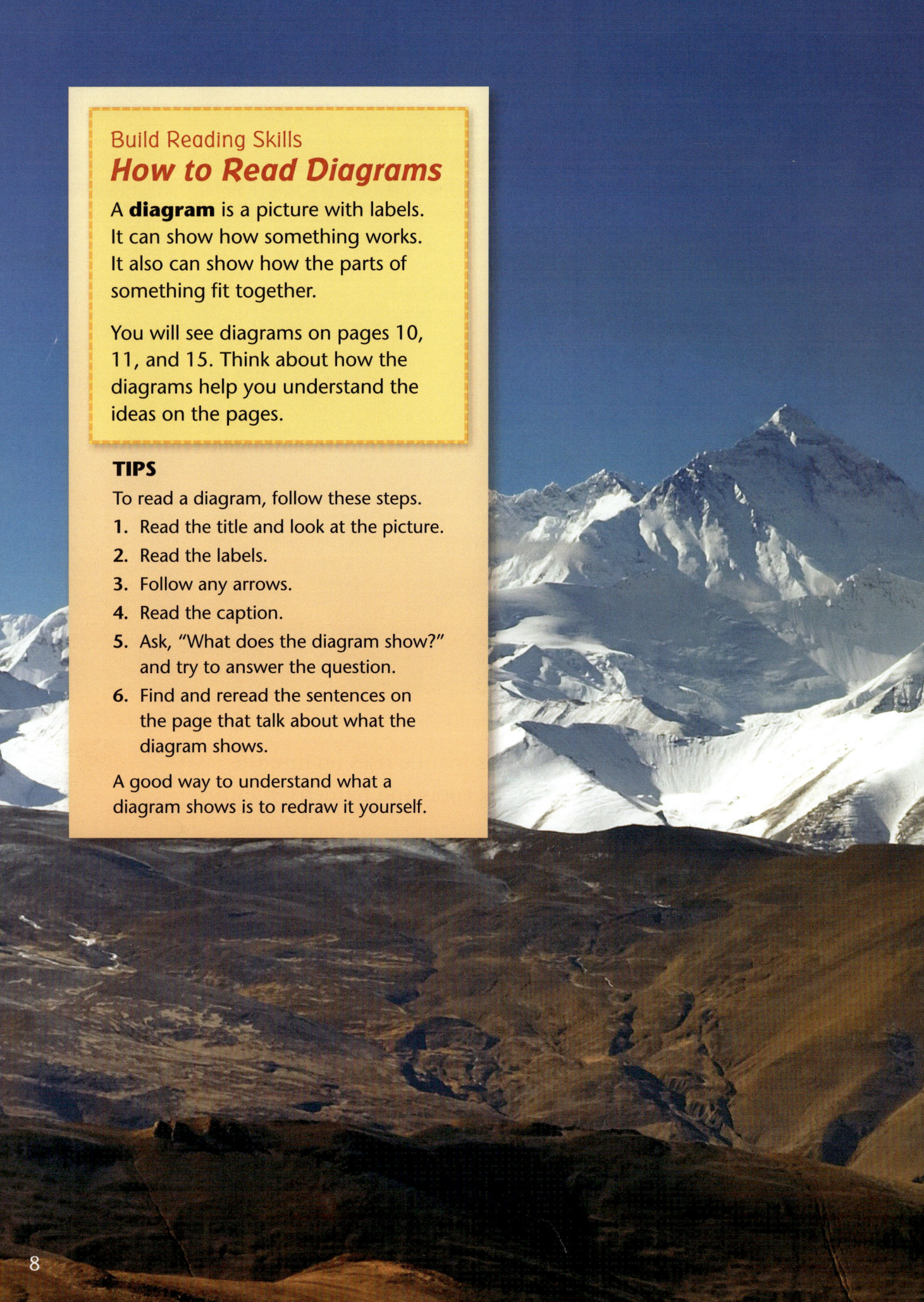

Build Reading Skills
How to Read Diagrams

A **diagram** is a picture with labels. It can show how something works. It also can show how the parts of something fit together.

You will see diagrams on pages 10, 11, and 15. Think about how the diagrams help you understand the ideas on the pages.

TIPS

To read a diagram, follow these steps.
1. Read the title and look at the picture.
2. Read the labels.
3. Follow any arrows.
4. Read the caption.
5. Ask, "What does the diagram show?" and try to answer the question.
6. Find and reread the sentences on the page that talk about what the diagram shows.

A good way to understand what a diagram shows is to redraw it yourself.

How Does Earth Change?

MAKE A CONNECTION
Mount Everest is the highest mountain in the world. It is in the Himalayas, a large group of mountains in Asia. How do you think these mountains formed?

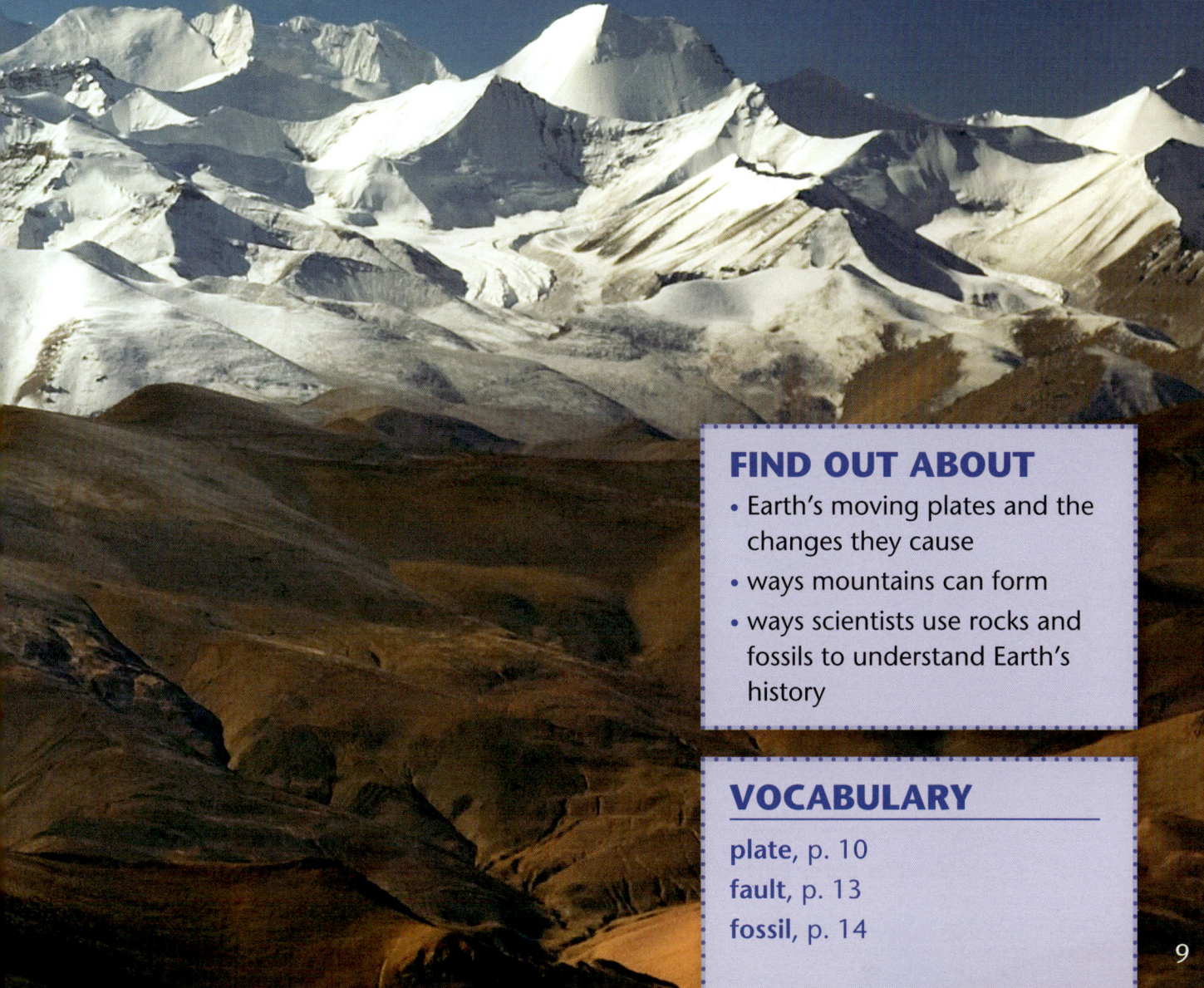

FIND OUT ABOUT
- Earth's moving plates and the changes they cause
- ways mountains can form
- ways scientists use rocks and fossils to understand Earth's history

VOCABULARY

plate, p. 10
fault, p. 13
fossil, p. 14

Earth's Plates

The outer part of Earth has moving pieces called plates. A **plate** is a thick piece of crust and part of the upper mantle. Earth has about 14 major plates. It also has some smaller plates.

Remember that some of the rock in the mantle flows very slowly. This movement causes the plates above to move slowly. Some plates move only about 2.5 centimeters (about 1 inch) per year. Other plates move as much as 15 centimeters (about 6 inches) per year.

Plate movements are always changing the land. Some changes caused by plate movements happen very slowly. Other changes happen very quickly. Important events happen at the places where plates meet. These places are called plate boundaries.

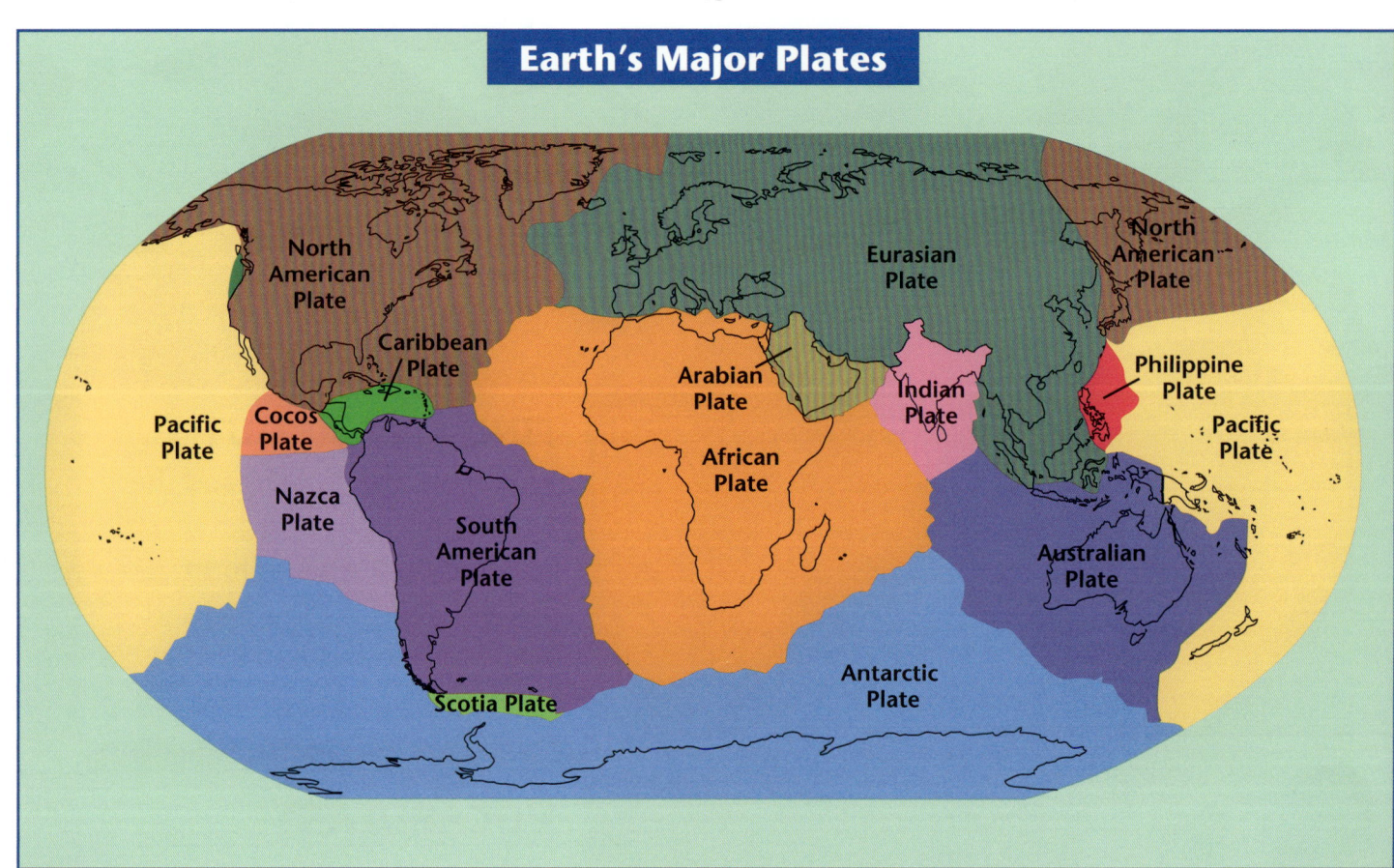

▲ The outer part of Earth has moving pieces called plates.

Important events happen at plate boundaries. ▼

Plate Boundaries

Plates slowly move toward each other and they crash.
The crust crumples and folds. Mountains form. This happens when two plates with continental crust meet.

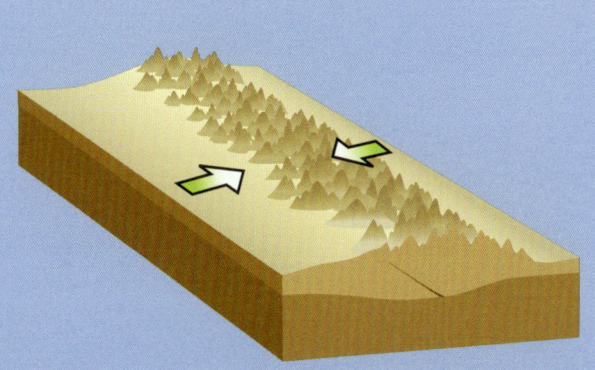

Plates slowly move toward each other. One plate sinks down under the other.
The crust of the sinking plate melts and is destroyed. A trench and volcanic mountains form. This happens when a plate with oceanic crust meets a plate with continental crust. It also happens when two plates with oceanic crust meet.

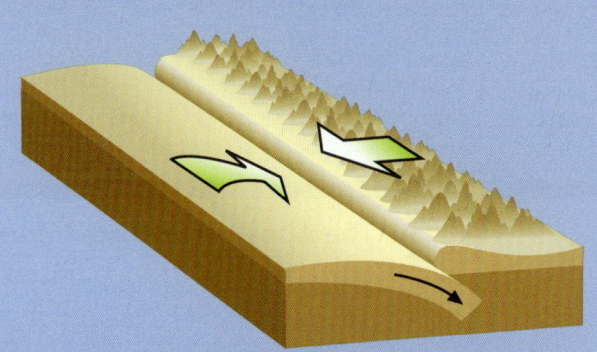

Plates slowly move apart.
Melted rock from the mantle pushes up between the plates. New crust is made as the melted rock cools and hardens. This forms mountains.

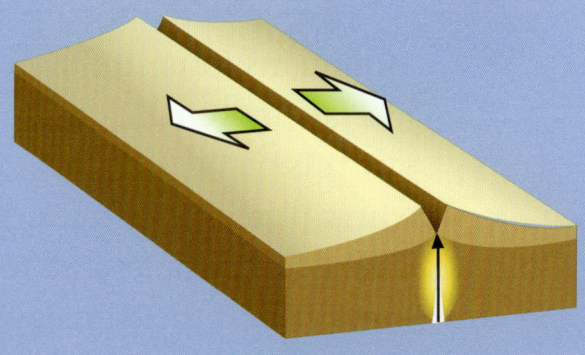

Plates slowly slide past each other.
Crust is not made or destroyed.

 What is a plate? Give an example of what can happen where two plates meet.

▲ The Appalachians are folded mountains. They formed along the eastern edge of the United States and Canada.

How Mountains Form

A group of mountains is called a mountain range. Mountains form in different ways.

Plate movements cause some mountains to form. You saw this on page 11. For example, folded mountains form where plates move toward each other and slowly crash. The crust crumples and folds. The Appalachians in North America and the Himalayas in Asia are folded mountains.

Volcanic mountains can form where plates are moving toward each other and one plate sinks under the other plate. Or they can form where plates are moving apart. In either case, melted rock comes to Earth's surface, cools, and hardens. The rock builds up over time. The mid-ocean ridge is a volcanic mountain range. The Cascades are also volcanic mountains. The Cascade range includes Mount Rainier and Mount St. Helens in the state of Washington. You will read more about volcanic mountains on pages 18–20.

Fault-block mountains form where large blocks of crust move up or down along a fault. A **fault** is a break in the crust. Many faults are at or near plate boundaries. But faults can be on other parts of a plate, too.

Dome mountains usually form far from plate boundaries. A dome mountain forms where melted rock pushes up on the crust above it. But the melted rock does not break through. It hardens below Earth's surface. This makes a mountain with a rounded top.

✓ What are two ways that mountains can form?

▲ The Grand Tetons are fault-block mountains in Wyoming.

▲ The Black Hills are dome mountains in South Dakota.

13

▲ Some rocks form in layers, with newer rock on top of older rock. But plate movements can fold, break, lift, and overturn rock.

Studying Earth

Scientists who study Earth are called geologists. Finding the ages of rocks helps geologists understand Earth's past.

Some rocks form in layers. Newer rock forms on top of older rock. Knowing this helps geologists tell the age of one rock compared with another. But plate movements and faults can fold, break, lift, and overturn rock. So the newest rock is not always found on top.

Geologists use fossils to find the age of some rocks. **Fossils** are the remains or traces of living things from long ago. Suppose a rock has a fossil of an animal in it. Scientists may know when that kind of animal was alive. The rock likely formed during that same time.

Geologists also can do laboratory tests to find the age of a rock. The oldest rocks found on Earth are about 4.3 billion years old. Scientists estimate that Earth is about 4.6 billion years old.

Earth's Moving Continents

Earth's continents were once joined together. Over time, they have moved to where they are now. In fact, the continents are still moving!

Some clues helped scientists discover that the continents moved. They saw that the shapes of South America and Africa seem to fit together like puzzle pieces. Also, the kinds of rocks found in South America match the rocks in Africa. And the same kinds of fossils have been found on both continents.

Plate movements cause the continents to move. Earth's plates include the crust of the continents. So, as the plates move, the continents move. Plate movements and other Earth processes happening today also happened in the past.

 What causes Earth's continents to move?

225 million years ago

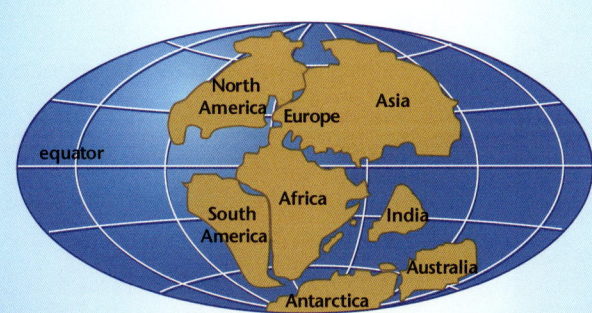
135 million years ago

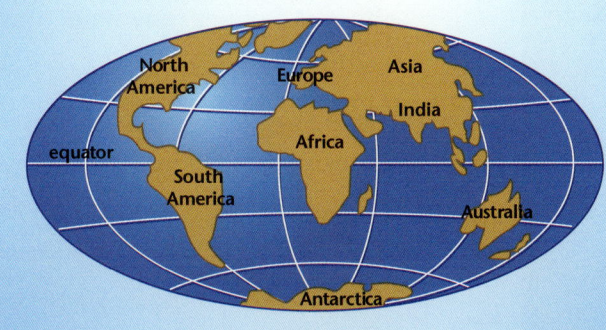
present day

Long ago, the continents were joined. They have moved apart over time. ▶

REFLECT ON READING
Look again at the diagrams on page 11 showing plate boundaries. Redraw one of the diagrams in your science notebook. Tell a partner what your drawing shows.

APPLY SCIENCE CONCEPTS
Choose a mountain you would like to visit. Read about it in books and on the Internet. In your science notebook, write the mountain's name and where it is on Earth. Then add one interesting fact about it.

15

Build Reading Skills
Cause and Effect

A **cause** is why something happens. An **effect** is what happens as a result of the cause.

You will read about volcanoes on pages 18–20. Think about the effects that an erupting volcano can have on Earth's land and air.

TIPS

Thinking about causes and effects can help you understand why things happen.

- To find effects, ask, "What happens?"
- To find causes, ask, "Why does this happen?"
- Look for signal words such as *cause*, *effect*, *affect*, *because*, *why*, *since*, *so*, and *as a result*.
- A cause may have more than one effect. An effect may have more than one cause.

A cause and effect chart can help you keep track of your ideas about why things happen.

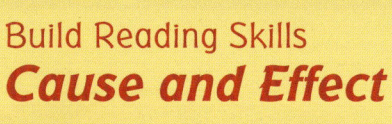

What Are Volcanoes and Earthquakes?

MAKE A CONNECTION
Lava from a volcano is hot, melted rock. It comes from deep inside Earth. What changes to Earth's surface do you think lava can cause?

FIND OUT ABOUT
- what happens when a volcano erupts
- three main kinds of volcanoes
- the cause of earthquakes
- measuring earthquakes

VOCABULARY

volcano, p. 18
magma, p. 18
lava, p. 18
earthquake, p. 21

Thousands of earthquakes happen every day. Most are too weak for people to feel. The amount of energy given off by an earthquake is called its magnitude.

A *seismograph* is a tool that keeps track of the vibrations from an earthquake. The record made by a seismograph is called a seismogram. Scientists study seismograms to learn

- the magnitude of an earthquake
- where an earthquake happened
- the time an earthquake started
- how long the earthquake lasted

The Richter scale is used to describe the magnitude of an earthquake. Earth has about 17 major earthquakes and 1 great earthquake each year.

▲ Seismographs keep track of the vibrations from earthquakes.

Richter Scale

Magnitude	Class
2.9 or less	micro
3.0–3.9	minor
4.0–4.9	light
5.0–5.9	moderate
6.0–6.9	strong
7.0–7.9	major
8.0 or greater	great

▲ The Richter scale helps us describe and compare the magnitudes of earthquakes.

▲ A strong earthquake can cause great damage to roads and buildings.

A strong earthquake can quickly cause great damage. The ground can crack. Buildings may be destroyed. Power, gas, or water lines may break. Fires may start because of this damage. Places where animals and plants live may be changed or ruined. An earthquake also can cause a landslide. Large amounts of rock and soil can slide downhill in seconds.

A strong earthquake on the ocean floor can cause a tsunami. A tsunami is a series of fast waves in the water. These waves become huge and cause great damage when they reach land.

 What causes an earthquake?

REFLECT ON READING
Make a cause and effect chart like the one on page 16. Write "erupting volcano" in the cause box. What effects can an erupting volcano have on Earth's land, air, and living things? Add these to the chart.

APPLY SCIENCE CONCEPTS
Do you think scientists can predict when an earthquake will happen? Write a paragraph in your science notebook telling your ideas.

Glossary

abyssal plain (uh-BIS-uhl PLAYN) a very wide, flat area of the deep-ocean floor **(6)**

continental rise (kon-tuh-NEN-tl RYZE) the underwater section of land that lies at the bottom of the continental slope, where the land starts to flatten out **(6)**

continental shelf (kon-tuh-NEN-tl SHELF) the gently sloping underwater edge of a continent **(6)**

continental slope (kon-tuh-NEN-tl SLOHP) the steeply sloping underwater section of land that lies between the continental shelf and the continental rise **(6)**

core (KOR) the center part of Earth, made mostly of metals; the outer core is liquid and the inner core is solid **(5)**

crust (KRUHST) the outer layer of Earth, made of solid rock **(4)**

earthquake (URTH-kwayk) vibrations of the ground that happen when rock below Earth's surface suddenly moves **(21)**

fault (FAWLT) a break in Earth's crust along which movement can take place **(13)**

fossil (FOS-uhl) the remains or traces of a living thing from long ago **(14)**

landform (LAND-form) a natural shape, or feature, on Earth's surface, such as a mountain, valley, plain, or plateau **(5)**

lava (LAH-vuh) magma that has reached Earth's surface **(18)**

magma (MAG-muh) hot, melted rock that is below Earth's surface **(18)**

mantle (MAN-tl) the layer of Earth between the crust and the core; some of the rock in the mantle flows very slowly because of the high temperature and pressure there **(4)**

plate (PLAYT) a thick, moving piece of Earth's crust and part of the upper mantle **(10)**

volcano (vol-KAY-noh) an opening, or vent, in Earth's crust through which lava, ash, and other materials erupt; also the mountain formed from these materials **(18)**